THE TESSELLATIONS FILE

40 BLACKLINE MASTERS WITH TEACHER'S NOTES

CHRIS DE CORDOVA

TARQUIN PUBLICATIONS

WHAT ARE TESSELLATIONS?

Broadly speaking, a tessellation is a design which is made up of repeated tiles. It is usually understood that these tiles are straight-sided polygons, but the word tessellation is also often used in a less strict sense to include tiles with curved-sides. The essential point is that the tiles should repeat and that no gaps should be left. Tessellations are also known as *plane tilings* because they could continue for ever in every direction and so cover the infinite plane. The best known exponent of plane tilings with curved edges is M.C.Escher. He discovered or invented a remarkable collection of interesting shapes which fit together in ingenious ways to cover the plane. A study of his work is a natural adjunct to the ideas suggested in this file.

These introductory pages explain how tessellations can be defined and how they can be classified into four main groups. They also offer some ideas of how easy it is to create an almost infinite variety of interesting variations by changing, blending and adapting the standard ones. The mathematical theory of regular polygons and the regular and semi-regular tessellations is given in some detail as it is fundamental to all of this work. However, the study of tessellations is valuable over a wide range of ages and abilities and the theory can be given up to any level the child is capable of understanding. Those children who are able to go on further than the others can be presented with some of these more advanced ideas to investigate in the form of a project. They often need very little guidance and usually find the possibilities absorbing.

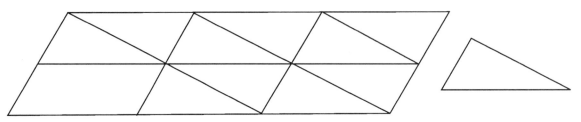

As part of the introduction to the subject, it is valuable to spend a little time making templates or cardboard tiles and drawing round them. This may seem a rather a laborious process and the finished results are seldom very pleasing as it is usually found that the tiles start to acquire rounded corners and after only a few repetitions fail to fit exactly. However, this process is an important first step. As soon as the principle of tiling and repetition is fully understood it is generally more satisfactory to move on to the use of accurate computer-drawn examples, such as the ones in this file.

In a classroom situation, one can suggest different sets of arbitrary rules for colouring the tiles. For instance, the use of only two, three or four colours. Alternatively no comment at all about the number of colours need be made and the different choices commented on later. In spite of quite a lot of evidence to the contrary most children have an in-built natural inclination towards order and symmetry and even when left to do as they please, will mostly choose a repetitive colour scheme. There is much to be learned about space-filling and geometry as they colour in the areas. The results of such a session can look stunning and when mounted and framed can form the basis of a fine exhibition for an open evening.

THE CLASSIFICATION OF TESSELLATIONS

Visually one tends to recognise tessellations by the shapes of the polygons which are present. For instance, one could describe a particular tessellation as being made up of squares or perhaps squares and isosceles triangles or perhaps even of squares, hexagons and dodecagons. However, such verbal descriptions are not really precise enough to classify tessellations properly or to be certain that all possible ones have been found. The most effective way of doing so is to use a symbolic method and the most convenient is known as the modified Schläfli symbol. It states, in numerical form, what happens at each node.

A node is the name given to the places where the corners of the polygons meet and the Schläfli symbol lists the order and number of sides of the polygons meeting at that node.

Tessellations can be broadly divided into four groups.

1. Regular tessellations where the only polygon is regular and identical.
2. Semi-regular tessellations where all the polygons are regular, but two or more are present.
3. Non-homogeneous tessellations where the polygons are regular, but not all nodes are identical.
4. Transformations and variations.

GROUP 1 REGULAR TESSELLATIONS

Since the angles at every node have to add up to 360°, it is apparent that there can only be three regular tessellations. If the only polygon present is to be regular and identical then it must be either

(a) an equilateral triangle,   (b) a square,   or   (c) a regular hexagon.

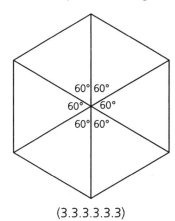
(3.3.3.3.3.3)

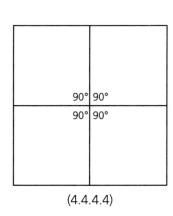
(4.4.4.4)

(6.6.6)

The diagram above is easy to understand and is an excellent way to introduce the Schläfli symbol for the first time. Pupils of almost any age can see that 6 x 60° = 360°, 4 x 90° = 360° and 6 x 120° = 360° and that there can be no other possibilities.

While this idea is being discussed it is well worth pointing out that if the sum of the angles meeting at a node is less than 360°, the polygons could form the vertex of a three-dimensional model.

The obvious illustration to use is that of three regular pentagons meeting at a point. Their angles add up to 324° (3 x 108°), so there would be a gap of 36° and this is not a node of a tessellation. Three pentagons will not lie flat, as three hexagons do, but could stand up to make a vertex of a dodecahedron.

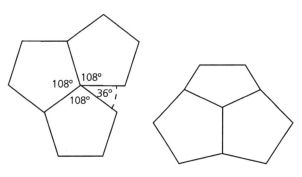

GROUP 2 SEMI-REGULAR TESSELLATIONS

When two or more regular polygons are present in a tessellation, there are many more possibilities and the construction of a table of the interior angles is an important preliminary step.

ANGLES OF REGULAR POLYGONS			
Name	Number of sides	Each exterior angle	Each interior angle
Triangle	3	360°/3 = 120°	60°
Square	4	360°/4 = 90°	90°
Pentagon	5	360°/5 = 72°	108°
Hexagon	6	360°/6 = 60°	120°
Heptagon	7	360°/7 = 51.428...°	128.571...°
Octagon	8	360°/8 = 45°	135°
Nonagon	9	360°/9 = 40°	140°
Decagon	10	360°/10 = 36°	144°
Dodecagon	12	360°/12 = 30°	150°
15-gon	15	360°/15 = 24°	156°
18-gon	18	360°/18 = 20°	160°
20-gon	20	360°/20 = 18°	162°
24-gon	24	360°/24 = 15°	165°
30-gon	30	360°/30 = 12°	168°
36-gon	36	360°/36 = 10°	170°
42-gon	42	360°/42 = 8.571...°	171.428...°
60-gon	60	360°/60 = 6°	174°
360-gon	360	360°/360 = 1°	179°

We can easily deduce that there must be at least three but not more than six polygons meeting at a node of a semi-regular tessellation. This is because the sum of the interior angles of the polygons meeting at the node must be 360° and this can only be achieved in a limited number of ways. It is clear that since no regular polygon has an interior angle of 180° or greater, there must always be more than two polygons present at a node. Also, because no regular polygon has an interior angle of less than 60°, there can never be more than six polygons present at a node. Hence there can only be 3, 4, 5, or 6 regular polygons meeting at a node of a semi-regular tessellation.

Using simple logic we can also deduce that there cannot be more than three different polygons meeting at any node. This must be true because the sum of the four smallest different interior angles is 378° (60° + 90° + 108° + 120°) and this is greater than 360°. Hence, if four, five or six polygons are to meet at a node then one or more of the smaller interior angles must be repeated.

With the aid of a table of possible interior angles such as the one overleaf and these restrictions, we can set out to find all the combinations which add up to 360°. An investigation like this makes an interesting class exercise and since there are 17 to find, it is achievable within a reasonable length of time. However, it would be a good class that discovers (3.7.42) without at least a hint of where to look!

This is the complete list together with their Schläfli classifications.

THREE REGULAR POLYGONS MEETING AT A NODE (There are ten possibilities)

120° + 120° + 120° (6.6.6)	108° + 108° + 144° (5.5.10)	90° + 120° + 150° (4.6.12)
90° + 135° + 135° (4.8.8)	90° + 108° + 162° (4.5.20)	60° + 144° + 156° (3.10.15)
60° + 150° + 150° (3.12.12)	60° + 140° + 160° (3.9.18)	60° + 135° + 165° (3.8.24)

60° + 128.571...° + 171.428...° (3.7.42)

FOUR REGULAR POLYGONS MEETING AT A NODE (There are four possibilities)

90° + 90° + 90° + 90° (4.4.4.4)   60° + 60° + 120° + 120° (3.3.6.6)
60° + 90° + 90° + 120° (3.4.4.6)   60° + 60° + 90° + 150° (3.3.4.12)

FIVE REGULAR POLYGONS MEETING AT A NODE (There are two possibilities)

60° + 60° + 60° + 60° + 120° (3.3.3.3.6)   60° + 60° + 60° + 90° + 90° (3.3.3.4.4)

SIX REGULAR POLYGONS MEETING AT A NODE (There is only one possibility)

60° + 60° + 60° + 60° + 60°+ 60° (3.3.3.3.3.3)

The list above was generated simply by considering the requirement that the angles at a node must add up to 360°. This must be satisfied, but it is not a condition which is sufficient to ensure that they all form tessellations. A little drawing and construction will probably be needed to convince most people that only some of them will tessellate the plane. The list must be reduced by eliminating those which do not. The three regular tessellations (3.3.3.3.3.3), (4.4.4.4) and (6.6.6) are included, so there remain 14 different ones to check.

For a tessellation to be considered semi-regular, the same polygons in the same order must occur at every node. There is no easy way to look at the symbol or at the angles and to be able to work out which will be tessellations and which will not. To convince yourself or a class, try (3.3.6.6) and (3.6.3.6). One makes a tessellation where all the nodes are the same and the other does not. It will also be found that two distinct semi-regular tessellations come from the single combination given above as (3.3.3.4.4). This is because the two arrangements (3.3.3.4.4) and (3.3.4.3.4) give different patterns, both of which meet all the conditions for a semi-regular tessellation. Finally, the combination (3.3.3.3.6) is an interesting one. It is treated as a single semi-regular tessellation but it can be drawn in two versions, each of which is the mirror image of the other.

When all of these tests have been applied to the seventeen combinations above, we are able to identify three regular and eight semi-regular tessellations. We can also be convinced that there are no others. All are included in this collection.

REGULAR TESSELLATIONS	SEMI-REGULAR TESSELLATIONS	
TESS^N 1 (3.3.3.3.3.3)	TESS^N 4 (3.6.3.6)	TESS^N 8 (3.3.3.4.4)
TESS^N 2 (4.4.4.4)	TESS^N 5 (4.8.8)	TESS^N 9 (3.3.4.3.4)
TESS^N 3 (6.6.6)	TESS^N 6 (3.4.6.4)	TESS^N 10 ⎱ mirror image versions
	TESS^N 7 (4.6.12)	TESS^N 11 ⎰ of (3.3.3.3.6)
		TESS^N 12 (3.12.12)

GROUP 3 **NON-HOMOGENEOUS TESSELLATIONS**

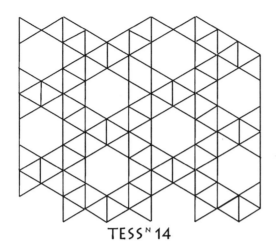

TESS^N 14

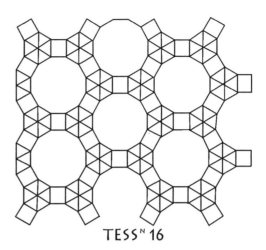

TESS^N 16

The word homogeneous means 'consisting of parts all of the same kind' and homogeneous tessellations are those where the same polyhedra occur in the same order at every node. Thus the regular and semi-regular tessellations so far discussed are all homogeneous. The phrase 'non-homogeneous' in work on tessellations might be thought to describe any other kind of tessellation but in fact it is used in a more restricted sense. It is used to describe tessellations of regular polygons but where the polygons are arranged so that within the design, there are several different types of node. Many people think that non-homogeneous tessellations are the most beautiful kind of tessellation, combining as they do, regularity with a measure of variety. Four examples are given (TESS^N 13 to TESS^N 16), but there are many more to be found.

They are often formed from and related to one of the regular or semi-regular tessellations. By making small changes to some of the polygons in one of the standard tessellations, a new and attractive design can be created. For instance, TESS^N 16 started with the semi-regular (4.6.12). Then all the hexagons were divided into six equilateral triangles. TESS^N 14 is derived directly from the regular tessellation (6.6.6), although it might not look like it at first sight. It can be very exciting indeed to discover a pattern within a pattern by experimenting in this way.

GROUP 4 **VARIATIONS ON A THEME**

Having started to modify tessellations with non-homogeneous tessellations it would be a shame not to encourage children to relax the rules still further and to find other ways of developing patterns. All the tessellations in this group are transformations of others in earlier groups.

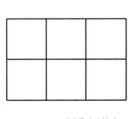

TESS^N 2

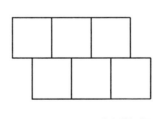

TESS^N 17

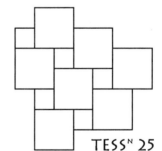

TESS^N 25

In the more formal tessellations, it was always required that all elements joined up edge to edge and corner to corner as in TESS^N 2. If this requirement is abandoned and if half slips are allowed, then designs such as TESS^N 17 become possible. TESS^N 25 shows the kind of design which can be produced if different side lengths are permitted. Some joins take place corner to corner and some do not.

  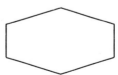

If we drop the requirement that polygons are regular, then many different designs are possible. We can approach this kind of tessellation by thinking of squashing, stretching and rotating the elements of a regular tessellation in a systematic way. For instance a regular hexagon can be distorted in various ways without losing its ability to tessellate the plane.

DUPLICATING AND MOVING A TESSELLATION

A rich source of variations can be found by photocopying a tessellation on to a clear acetate sheet and then placing this new sheet over the original. It can be made to match exactly and then consciously moved by varying amounts to generate an infinity of new and complex patterns. By stopping when certain nodes come on top of others and when certain lines coincide, the pattern can be seen to magically simplify and thus suggest drawable tessellations.

This is a particularly effective demonstration to do on an overhead projector where both copies of the tessellation are on clear acetate sheets. One is taped to the base and the other moved across it.

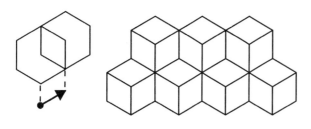

This tessellation gives the impression a three-dimensional set of piled up blocks, but it can also be seen as a simple variation of the design (6.6.6).

It can be seen or generated in two different ways, either as three vertices of each hexagon being joined to its centre or as a duplicate set of hexagons superimposed with a half-slip in the direction of the arrow.

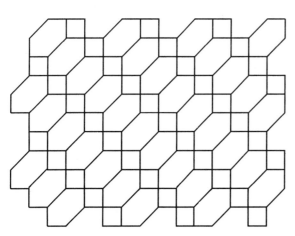

This tessellation is also a variation on (6.6.6). At first sight it appears to be a combination of squares and squashed hexagons.

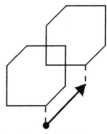

Another way of looking at it is in terms of hexagons of this shape, duplicated and slipped as in the previous variation. This time there is no impression of three-dimensionality.

MODIFYING SEMI-REGULAR TESSELLATIONS

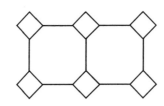

TESS^N 5

These patterns are very well known and easily recognised, as different versions of it are widely used on vinyl for kitchens and bathrooms. TESS^N 5 is the semi-regular tessellation (4.8.8) where the octagons and squares have the same edge length. The variations on the right show what happens when the proportions are increased and reduced. Any proportions give a tessellation and notice how, at both limits, when either the square or the octagon reduces to zero, it simplifies to the regular tessellation (4.4.4.4).

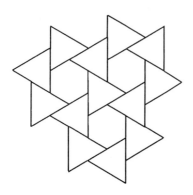

These two designs are both variations on the semi-regular tessellation (3.6.3.6).

It can be modified into non-homogeneous tessellations of very different appearance, according to which of the two shapes is increased relative to the other.

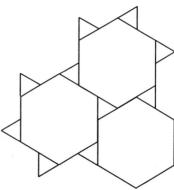

CHANGING THE LENGTHS OF EDGES

The previous section showed how interesting non-homogeneous tessellations could be made with pairs of squares and octagons and also with pairs of hexagons and equilateral triangles.

It is also interesting to experiment with pairs of different sizes of a single shape. Try it with pairs of squares in any proportion and also with rectangles, parallelograms and scalene triangles.

You will find that it does not work with hexagons

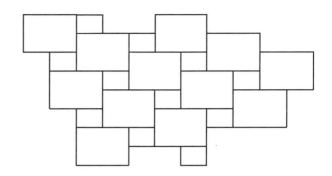

MODIFYING THE ELEMENTS OF REGULAR TESSELLATIONS

by translation

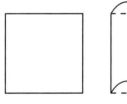  

by rotation

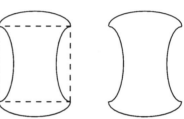

Starting with any of the three regular tessellations, one can take a piece out of one side and add it to the opposite side. It is obvious that such transformations must produce shapes which will tessellate and with some imagination and experiment, very interesting designs can be produced.

Another good starting point for this approach is a tessellation of parallelograms.

Tessellations where all the edges are curved can be be very easily devised using simple transformations like these. It can be quite hard to see that a design like this is based on an underlying square grid.

If you take a shape from the sides and add it to the top and bottom, them it will tessellate, but with an interesting rotational symmetry. In the illustration above, the line segments have been rotated anti-clockwise by 90° about the vertices of the square.

MODIFYING ALL EDGES SYMMETRICALLY

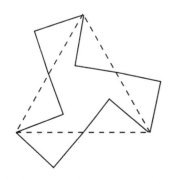 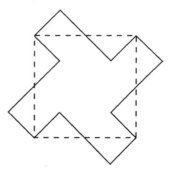 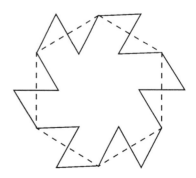

A whole range of interesting shapes which tessellate can be created by modifying each edge of a regular tessellation in a way which has rotational symmetry about its mid-point. For every piece which is added to one half of the edge, a corresponding piece has to be taken away from the other half.

TESSELLATING THE LETTERS OF THE ALPHABET

An interesting project is to try to draw some letters of
the alphabet so that they will tessellate. The best
way to start is to sketch them on squared paper.
Some tolerance is needed with shapes and proportions but satisfactory results can often be obtained.

Sometimes a single letter, like this F, will tessellate
but it can be hard to find the pattern and it may take
several attempts before an answer is found.

If it can be achieved, it is most pleasing to be able to
create a monogram from someone's initials and then
tessellate it.

THE DUAL OF A TESSELLATION

To draw the dual of a tessellation, join the centre of every shape to the centres of its neighbours. By the
nature of the method, this will always produces a tessellation, generally a different one. The result may well
turn out to be less regular than one might perhaps expect.

It is especially recommended to try:

(3.6.3.6)	- produces a six-pointed star and/or diamond tessellation
(3.3.3.3.6)	- produces a tessellation of irregular pentagons
(3.12.12)	- fascinating design of six-pointed stars in hexagons. Very difficult to colour.
(6.6.6)	- produces (3,3,3,3,3,3)
(4.8.8)	- produces diagonally divided squares
(4.6.12)	- produces hexagons divided into 12 right-angled triangles

PENROSE TILINGS

Having demonstrated that regular pentagons will not
tessellate it is very useful to be able to demonstrate
that there are tessellations that have five-fold symmetry. Penrose tilings based on 'thick' and 'thin' rhom-
buses do cover the plane and do have five-fold symmetry in parts. This is known as 'quasi-symmetry'
because remarkably the plane has no overall lines of
symmetry. If a duplicate of the tessellation is made
on an infinite sheet of clear plastic as suggested on
page 6, there is nowhere else at all to which it could
be moved and replaced on the original so that the
pattern would exactly match.

This is in complete contrast to all the other tessellations in this collection where the patterns match
when replaced on the original at an infinite number
of different locations.

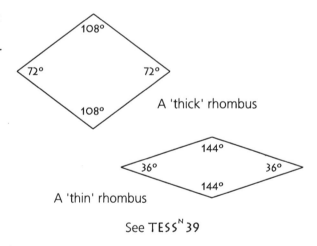

A 'thick' rhombus

A 'thin' rhombus

See TESS[N] 39

BIBLIOGRAPHY

SMP Book B - Teachers' Book - Cambridge University Press.
Mathematical Games and Pastimes - Kraitchik - Allen & Unwin.
Mathematical Snapshots - Steinhaus - Oxford University Press, New York.
Tessellations - J. Mold - Cambridge University Press.
Introduction to Tessellations - Seymour & Britton - Dale Seymour Publications.
Mathematical Models - Cundy & Rollett - Tarquin Publications.
Tilings & Patterns, an introduction - Grünbaum & Shephard - W. H. Freeman.
M.C.Escher, Visions of Symmetry - Schattschneider - W. H. Freeman

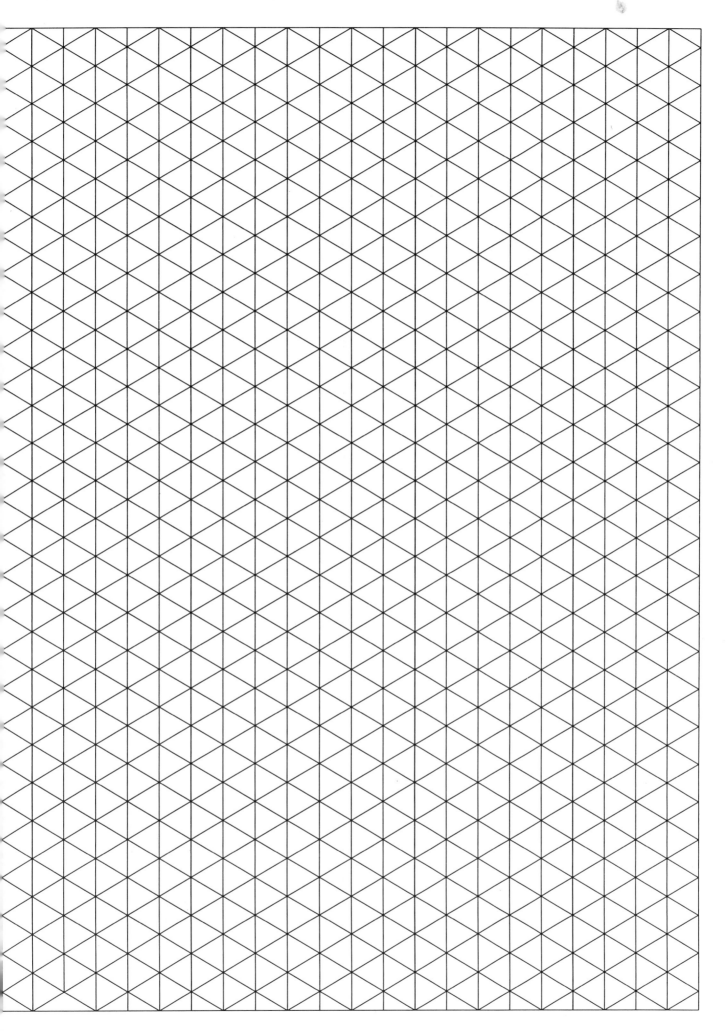

TESSELLATION **1**

TESSELLATION 2

TESSELLATION 3

TESSELLATION **4**

TESSELLATION 5

TESSELLATION 6

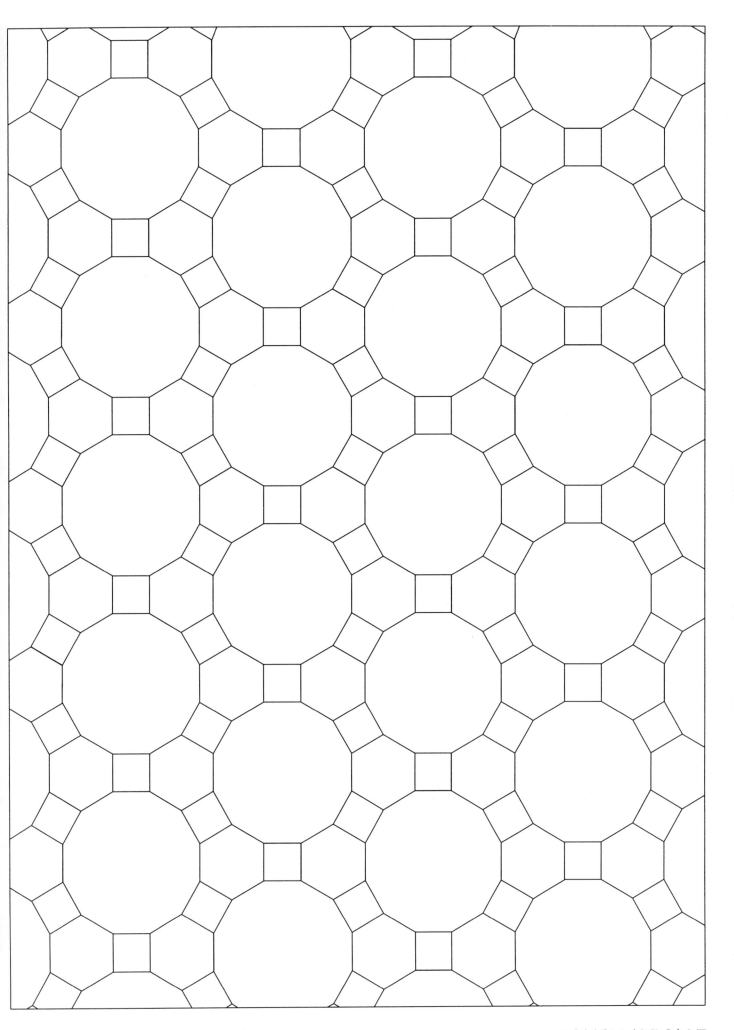

TESSELLATION **8**

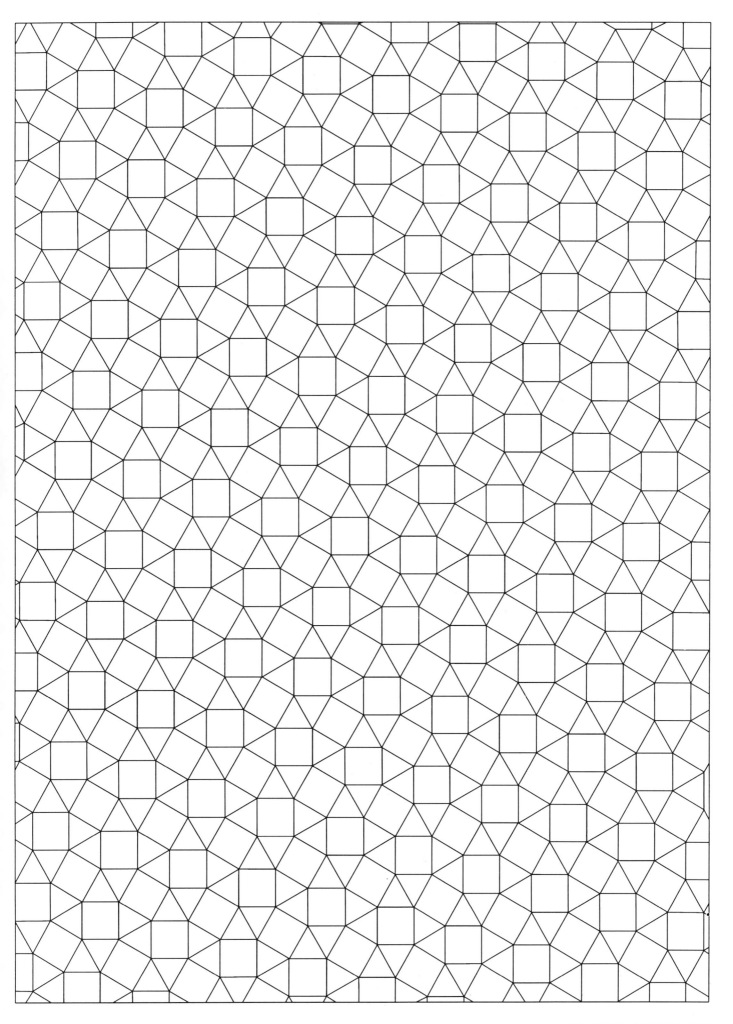

TESSELLATION 10

TESSELLATION **12**

TESSELLATION **13**

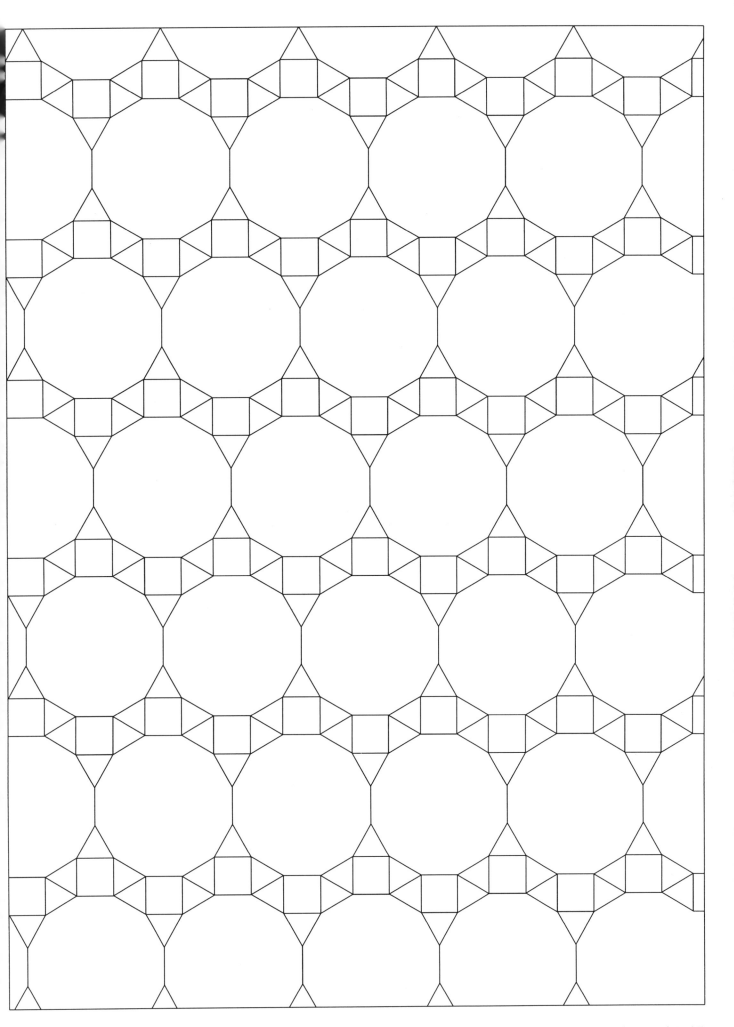

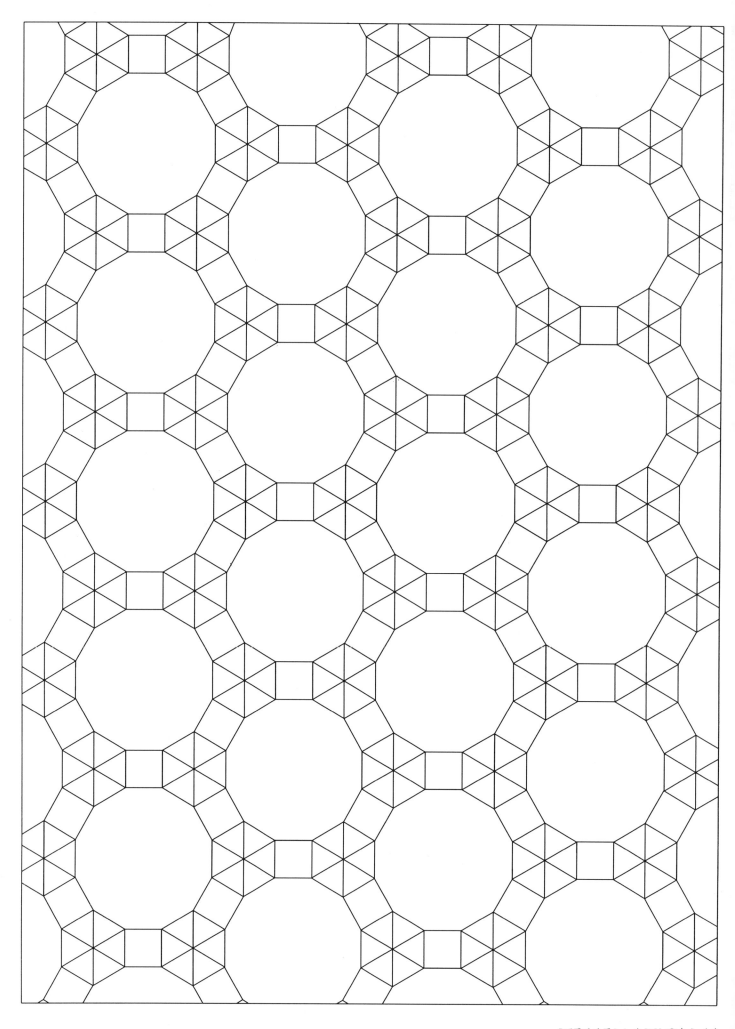

TESSELLATION **16**

TESSELLATION **18**

TESSELLATION **20**

TESSELLATION **21**

TESSELLATION 23

TESSELLATION 24

TESSELLATION **25**

TESSELLATION 26

TESSELLATION **27**

TESSELLATION **28**

TESSELLATION **30**

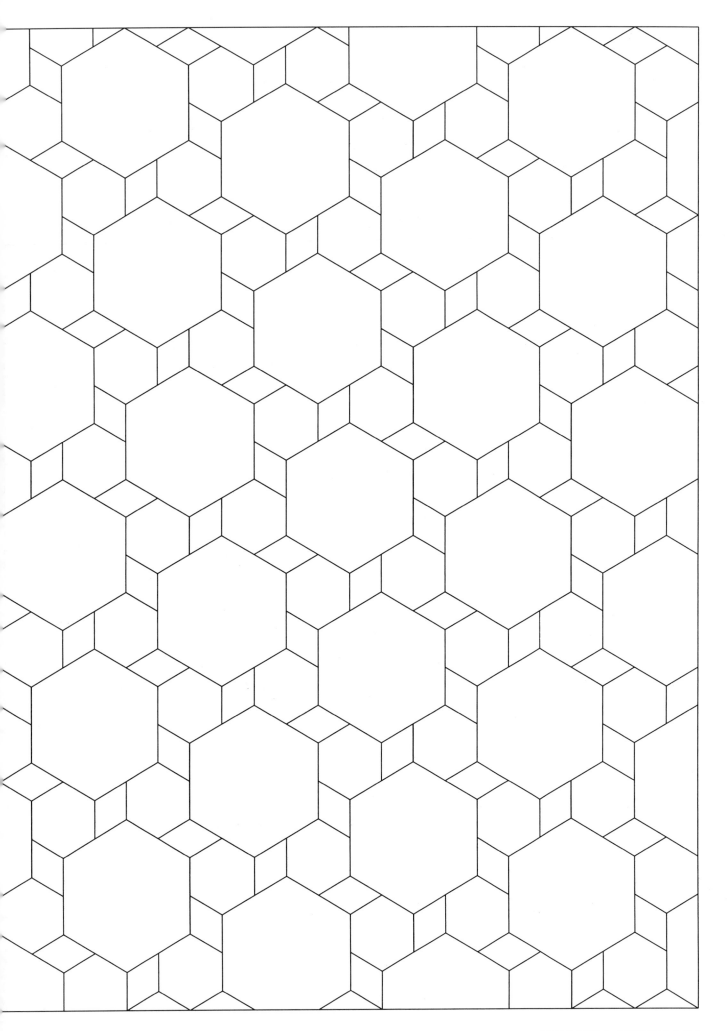

TESSELLATION 33

TESSELLATION **34**

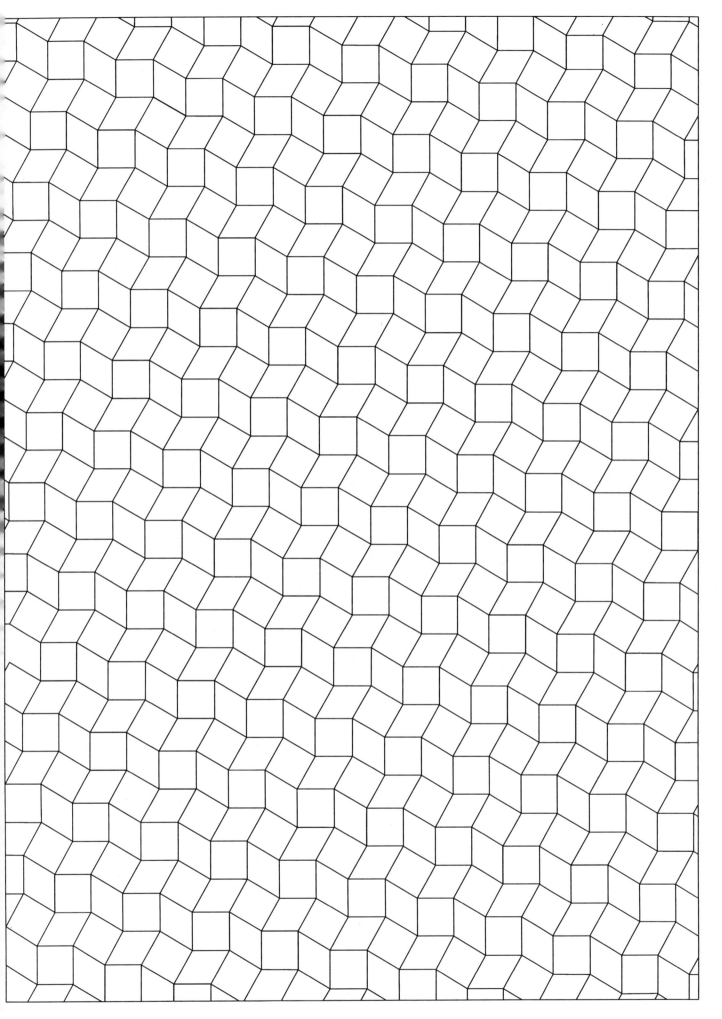

TESSELLATION 38

TESSELLATION 39

TESSELLATION 40